VOLUME 81

THE ECUATION OF THE UNIVERSE

THE UNIVERSE WAS FORMED OUT OF NOTHING BY THE MOVEMENT OF AN ALMATRINO

FIRST EDITION

Carlos L Partidas

Legal deposit number: MI2022000631

ISBN: 979 8370 8808 89

SAPI INTELLECTUAL PROPERTY REGISTRATION: NO. 8074 OF THE COMPENDIUM THE CHEMISTRY OF DISEASES BOLIVARIAN REPUBLIC OF VENEZUELA, 07/05/2010

DEDICATION

IN MEMORY OF THE ITALIAN SCIENTIST GALILEO GALILEI. WITH THE FIRST SCIENTIFIC INSTRUMENT IN HISTORY IN THE FORM OF A SMALL TELESCOPE, GALILEO GALILEI WAS ABLE TO DEMONSTRATE THAT THE CENTRE OF THE UNIVERSE WAS NEITHER THE EARTH NOR THE SUN. IT WAS A SCIENTIFIC EVENT THAT CHANGED MANKIND'S CONCEPTION OF THE CREATION OF THE UNIVERSE

TABLE OF CONTENTS

ACKNOWLEDGEMENT

TO THE RELATIVITY THEORIES OF ALBERT EINSTEIN AND MILEVA MARIĆ; AND, THE BIG BANG THEORY OF THE BELGIAN REVEREND GEORGES LEMAÎTRE. THESE TWO THEORIES REACH THEIR FINAL POINT IN SCIENTIFIC HISTORY, WITH THE EQUATION THAT EXPLAINS HOW THE UNIVERSE WAS FORMED FROM NOTHING

Chapter 1

THE INFINITESIMAL UNIVERSE

Events do not repeat themselves; for, the Universe expands exponentially and chaotically to where there is nothing. The Universe came into being by the movement of the smallest amount of energy that can fit in our mind. The Universe is imploding into nothingness. We can say that, the Universe is an energy system that was initiated by the movement of the smallest amount of energy that began to move in the centre of nothingness. Nothingness is not infinite; nothingness is the same size as the Universe. But, if we want to give a limit to the size of nothingness, nothingness grows expansively as the Universe expands. The Universe will not stop growing, because the expansion of the Universe is towards nothingness.

In nothingness there is nothing; however, we have to define the minimum amount of energy that formed in nothingness in order to connect nothingness with the Universe, and thus to describe how the Universe began to form out of nothingness.

Electronic energy arose by the movement of the minimum amount of electronic energy; and electronic matter was formed by the integration of electronic energy. Magnetic energy arose by the movement of electronic energy; and, magnetic mass was formed by the integration of magnetic energy.

Thus, everything that exists, that is, any physical system that is made up of electronic matter, including the bodies of living beings and spirits, are all formed as a consequence of the energy that is produced by the movement of the Universe.

The electronic matter that is produced in the Universe is not conscious of its existence; whereas the magnetic mass is the conscious part of the Universe; that is to say, the magnetic mass that forms the spirits is conscious of itself.

Thus, electronic matter and magnetic mass form an inseparable whole out of the Universe that is being formed out of nothingness.

On Earth, the magnetic mass of the spirit is the energy that drives the electronic matter of the physical body. But on Earth, the electronic matter of the physical body is believed to be the actual life form; therefore, the value of life is given only to the physical form of a human being. But any organism that moves its body by means of its magnetic mass is in reality a living being that also has the right to exist.

However, because of this ignorance of the origin of the life form, human beings are destroying the life of other living beings on planet Earth.

This difference is due to the fact that not everyone can see the energetic image of a spirit. The physical body of the earth and the physical body of a human being were formed by the integration of electronic energy; whereas, the magnetic mass of the spirit was formed by the integration of magnetic energy.

Electronic energy and magnetic energy are two different kinds of energies; therefore, these two kinds of energies form substances that are equally different. The difference is due to the fact that electronic matter is made up of spatially bound nuclei and electrons, whereas the magnetic mass of spirits has no nuclei or electrons; therefore, the magnetic mass of spirits is a more stable energetic pattern than the more stable electronic matter.

Electronic matter is made up of a kind of integration that can change over time, since the stability of the substance made up of electronic matter will depend on the arrangement of nuclei and electrons, i.e. on the compensation that occurs due to the more stable integration of electronic charges.

Whereas, the magnetic mass of a spirit has no nuclei or electrons; therefore, the magnetic mass of the spirit has no electronic charges. Therefore, the spirit cannot be modified or destroyed over time. We can say that the spirit is made up of the most stable substance that can exist.

Therefore, the magnetic mass of the spirit can only ride on the electronic matter of a physical body to conduct it, but without integrating with the physical body. Magnetic energy was formed by the movement of electronic energy; so, that, having no nuclei, electrons or electronic charges, the magnetic mass of the spirit will not be able to merge with the electronic matter of the physical body.

To the spirit, the electronic matter is transparent; to the spirit it is as if the electronic matter does not exist. For example, the spirit can pass through any kind of electronic matter without interacting, because the spirit has no electronic charges. The magnetic mass of spirit in relation to electronic matter would be like putting a sieve through water.

When we refer to magnetic energy, this is the holographic energy stamp that leads to the electronic body of any living being; that is, it can be the body of a tadpole, a fish, a whale, a virus, a bacterium, a tiger, a monkey, a cat, a bull, a cow, a goat, a chicken, a butterfly, a jaguar, a snake, a human being, etc.

But we have already said that the confusion on Earth is due to the fact that we do not see the energy of the spirit; for we can only see the physical body of any living being, or the physical body of any lifeless object, because these bodies are made up of electronic matter that can change with time. Whereas, the energy that drives the body of a living being we do not see, because it is a different energy; but, the energy of the spirit we have to classify as energy. To the driving stamp of the physical body we must for the time being call energy, until we can establish a name for the more stable substance that forms a spirit. In reality, the magnetic mass of the spirit is the holographic stamp that drives the physical body of living electronic matter.

The electrons and nuclei arrange themselves or change their spatial position; that is to say, the negative electrons and positive nuclei adjust their electronic charges according to the difference in integrating energy. This electronic arrangement will culminate when the electronic configuration formed is the most stable. For example, a stone, sand, a slag, the bark of a tree, etc.

These variations of electronic charges can be the hydrogen bridges that form between the four molecules in DNA: adenine, thymine, guanine and cytokine, which give an identity in the form of a genetic code to every body of every living thing on Earth: say, from a virus to a human being.

The actual form of life is the magnetic mass of the spirit, so that the spirit that has developed its reasoning the most is the one that is the highest on the evolutionary scale of existence. For example, a virus or a bacterium is a living being, but it is not conscious of its existence. Whereas the human being, by the knowledge he or she has acquired, is on a higher scale of existence.

Although there are human beings who function without being aware of their existence; for these human beings have not been awakened to the origin of their existence or to the state of consciousness. So we must delve deep into the consciousness of the unconscious human being in order to awaken that state of consciousness, so that all living beings can evolve. For all human beings to be placed on a higher degree of their evolutionary ladder.

Therefore, a pet or any animal may be below the scale of a human being's awareness, but a living being other than a human being expresses its feeling through instinct.

For example, a cow may not be aware of its existence; but a cow, a pig, a chicken, a deer or a sheep has feelings; therefore, killing them to eat the flesh of their bodies is only done by human beings who have failed to awaken the state of consciousness.

To kill a brother of the Universe who is alive in order to eat him, or to trade the flesh of its body, or to shoot him in the head from a distance in order to see him fall, is one of the most abominable acts of the unconscious human being. It is an aberrant act committed by the human being who is not aware of himself, of the existence of other beings, nor of the existence of the Universe.

The Universe is not aware of its existence; for the conscious part of the Universe is the magnetic mass of all the spirits that exist in the Universe. The conscious part of the Universe is the magnetic mass that forms the spirits of all living beings; but, on Earth, unconscious human beings assume that the only existence of life in the Universe is the physical bodies of human beings.

The magnetic mass of a spirit does not contain electronic matter; therefore, earthlings do not consider the existence of the holographic imprint of the magnetic mass of a spirit.

The microscopic arrangement of electronic charges has been going on for billions of years, and this process of adaptation will continue to happen in this invisible form. This arrangement of electrons and nuclei over billions of years is what has created the perfection that we notice at this moment of existence on Earth.

But, finding no explanation for this electronically occurring perfection, most people attribute it to a perfect being. But, in reality, that all that exists and will exist, we owe to the electronic arrangement that forms the generation of energy that emanates from the movement of the Universe.

So, everything that exists in the Universe is a consequence of the movement of the Universe; therefore, all living beings are brothers and sisters; both from the genetic point of view and from the energetic point of view, because everything that exists and what will exist in the Universe, is a consequence of the movement of the Universe. So we are all sons of the great Universe.

In fact, that we are all the Universe, because the conscious part of the Universe is the magnetic mass of the spirits. The spirits must have been formed in the different stages of energy; that is, in different energetic levels.

For example, in the beginning, or when the Universe was infinitesimal in size, infinitesimal electronic energy was formed; and by the motion of infinitesimal electronic energy, infinitesimal magnetic energy was generated. The rate of spin of the infinitesimal electronic energy reached an infinite value with respect to the infinitesimal size of the Universe; and the first quantity of infinitesimal electronic matter in the Universe began to form; that is, m_0 in the energy equation of the Universe $Ev=m_0C^3$. This is the energy equation that explains more clearly how the Universe was formed, because with this equation we can know at any time at what rotational speed the electronic energy is converted into electronic matter.

At the beginning, or when the Universe was infinitesimal in size, the infinitesimal magnetic energy generated was spontaneously integrated with other infinitesimal magnetic energy rotating in the same direction (↑↑ or ↓↓), and the first positive (↑↑) and negative (↓↑) quantity of infinitesimal magnetic mass was formed.

The infinitesimal magnetic mass and the infinitesimal electronic matter were spatially integrated and the first infinitesimal bodies were formed. For example, the body that forms the infinitesimal magnetic mass of a virus.

In this way, the different energy levels were formed, up to the size of the Universe as we know it today. But, the perfection by the arrangement of the electronic charges, it seems to us that it was someone who created the Universe. The process of electronic fine-tuning of this system of the Universe has been going on microscopically for billions of years, and will continue to go on for billions of years more; but it will be impossible for a human being to see the evolutionary process of the Universe.

A human being travels through space at zero speed relative to the Earth, and the Earth travels through space at 28 kilometres per second relative to the Sun.

In other words, we travel on the Earth with zero velocity relative to the Earth; therefore, it seems to us that the Universe is static or that the Universe does not move. Some human beings consider that all this perfection was produced by an invisible being similar to the human being. However, in order to create the Universe, we would have to imagine that this creator being is outside the Universe; that is to say, in the nothingness; but it is impossible for someone to exist in the nothingness because in the nothingness there is nothing.

In twice the present time, i.e. when the age of the Universe reaches 27.6 billion years, the Universe will still be creating itself and we will have a physical configuration different from the present configuration of the Universe. But, the magnetic mass of the spirits will be there, and will be able to contemplate the new events that are happening in the Universe, the new galaxies that will form new suns, the new living things, because the Universe will not stop growing as long as the Universe is in motion.

If we continue as we are going, the Earth will always be there in the Universe as a celestial body; but, perhaps the unconscious human being will destroy life on Earth before that time. Because unconscious humans consider the life of animals as an economic activity. Or the unconscious human being seeks his eternity in the transhumanism of the physical body, which is impossible, because the only eternal thing in the human being is the magnetic mass of the spirit. It will be impossible to immunise the magnetic mass of a spirit.

A fable from a philosophical point of view makes sense; therefore, human thinking was filled with convincing fantasies, when the explanation of how the Universe was formed was a scientific vacuum.

Until Galileo Galilei appeared with a scientific instrument represented by a small telescope to observe the Universe. Galileo Galilei realised that the sky does not exist, and that the centre of the Universe was not the Earth or the Sun; but, this demonstration by Galileo Galilei made the top of the Catholic church uncomfortable or those who were convinced of the philosophical ideas of Claudius Ptolemy.

However, the real or scientific idea of the formation of the energy of the Universe began with the conviction represented by Galileo Galilei's telescope.

The high energy was produced by the movement of the smallest amount of energy imaginable, which we have to define as an almatrino.

The heat generated in the infinitesimal Universe began to diminish when the electronic and magnetic energies were spontaneously integrated: the electronic energy was integrated and electronic matter was formed, and the magnetic energy was integrated and the magnetic mass of a spirit was formed.

Thus the different electronic and magnetic classes and bodies were formed, as the Universe expanded; for the expansion of the Universe is into nothingness.

In this way, space, the different physical electronic bodies and the different and distinct types of male and female magnetic beings were formed in this period of 13.8 billion years. On Earth, the physical body of a male and female being will be able to integrate in a spatial manner to form other functional beings.

Physical changes will continue to occur in the Universe, because the Universe, as it expands, seeks a state of minimum energy; that is, an energetic condition, where the amount of energy is less. But, as the Universe moves, it generates more energy; therefore, the Universe will not reach an end point, i.e., the Universe will not reach a point where it is in thermal equilibrium.

Chapter 2

ENERGY LEVELS

If we want to know what was there before the Universe was formed, we can consider it mathematically, to show that before the Universe was created, nothing existed in the nothingness.

Before the Universe was formed, we can write virtually that 0/0=<(0)>. It means that, the point zero of the Universe was inside a zero. We can write it in this virtual way, because: 0/0=0, 0/1=0, 0/2=0, 0/3=0 ...1000/0=0. It means: 0/0=0, 0/0=1, 0/0=2, 0/0=3 ...0/0=1000. But, 0≠1, 0≠2, 0≠3 and 0≠1000.

This inconsistency of numbers is known as the mathematical paradox of division by zero; which, can be resolved if we consider the numbers virtually; since, we can include the value of any number as a virtual expression of the form: <(n)>. In this way, the value 'n' has a virtual value before the point zero. That is, any numerical value can be written virtually. For example: 0/0=<(3)> but the value is not 3, but the value 3 is inside the value zero. This paradox of division by zero was solved by the young Venezuelan Ramsés Cornieles.

It means that, before the point zero in the nothing there was nothing, even if we do it in a virtual way, because we can go backwards to a point before the point zero of the Universe.

In an instant, the smallest energy that can fit in our mind was formed; and because it was energy, this minimum amount of energy started to move, and the formation of the Universe began at point zero.

But, we must keep in mind that the Universe is an energy system. So, this minimum amount of energy that was formed in the nothingness is what we have defined as an almatrino. So, we can connect the nothingness with the Universe in a mathematical way; which is what will lead us to a scientific reasoning.

An almatrino contains no electronic charge or mass; an almatrino is just energy in motion.

Similarly, the Austrian-born theoretical physicist Wolfgang Ernst Pauli had to resort to the concept of energy in order to balance the amount of energy missing from beta decay. Thus, Wolfgang Pauli's particle could only contain energy, i.e. this particle proposed by Pauli could contain neither charge nor electronic mass. However, at the time of Wolfgang Pauli, the existence of a particle without mass and electronic charge could not be understood. Therefore, Wolfgang Pauli said in a lecture:

" ... I have done something foolhardy, for I have proposed a particle that cannot be detected".

Wolfgang Pauli called this imaginary particle, which had neither charge nor mass, the neutron; however, the Italian physicist Enrico Fermi suggested to Wolfgang Pauli that he should call this particle the neutrino, because the neutron already existed. Finally, the existence of a neutrino was experimentally detected. However, an almatrino is smaller than a neutrino, so an almatrino cannot be detected experimentally.

An almatrino is an infinitesimal amount of energy that started to move at the initial point of the Universe, and began to generate an infinitesimal Universe.

At the first instant of motion, an almatrino could contain neither matter nor electronic charge; it had only one pole, since matter and electronic charge are formed by the speed of the spin. The monopole was proposed by the British mathematician and physicist Paul Dirac.

An almatrino has to rotate around itself in order to travel through an elliptical orbit.

The idea of a particle rotating on itself was proposed by the German theoretical physicist Ralph Kronig. However, Ralph Kronig's idea caused Wolfgang Pauli some irony. Thus, in a letter Wolfgang Pauli says to Ralph Kronig:

" ...a particle cannot rotate around itself, because this assumption would violate Albert Einstein's law of relativity".

Ralph Kronig then retracted his proposal in the face of the scientific prestige of Albert Einstein and Wolfgang Pauli.

Then, Wolfgang Pauli realises that Ralph Kronig was right; but, if you divide the energy value by 2. If you divide the energy value by 2, the spin of a particle around itself proposed by Ralph Kronig does not violate Albert Einstein's theory of relativity.

In fact, there is neither Albert Einstein's relativity nor Wolfgang Pauli's exclusion principle. Regarding the Pauli exclusion principle, what exists is a probability of energy levels. As for Albert Einstein's relativity, all variables in the Universe are absolute, once we know what the point zero of the Universe is.

There are also no quantum levels, but energy levels of a probability. The probability at the first energy level is 2 for event 1, i.e. +(½) and -(½).

That is, if in the same energy level, an electronic particle has an upward-flowing energy, the next particle to be formed must necessarily have a downward-flowing electronic energy, so that the two particles can exist in the same energy level.

This spin probability of a particle can be explained by fermions and bosons. So, in the same energy level we will have 2 fermions; whose electronic energy flows in the same direction. For example, upwards. We call these fermions whose electronic energy flows upwards positive electronic energy.

The electronic energy flowing upwards generates a magnetic energy that rotates from left to right, and we call it positive magnetic energy.

A boson is the energy probability that integrates 2 fermions whose electronic energy is flowing in the same direction. For example, at level 1, a boson is the integration probability of the fermions +(1/2) and +(1/2) or (↑↑). The sum of these 2 probabilities is 1. The other probability of integration of a boson is: -(1/2) + [-(1/2)] or (↓↓); this integration is also 1. That is, it is not necessary to put a mathematical sign on a boson, because the integration is only a probability.

Thus, at energy level 1, we will have sequentially the probabilities of motion of 5 alternately rotating almatrinos: +1/2, -1/2, +1/2, -1/2, +1/2 and +1/2; or (↑↓↑↓↑).

The electronic energy of the positive fermions +1/2 and +/1/2 or (↑↑) spontaneously integrate and the positive electronic matter of the electronic nuclei is formed.

If what we want is to compare this movement with an observable fact, two tornadoes that are rotating in the same direction (→→ or ←←) unite spontaneously and a single tornado will be formed; but, if the 2 tornadoes are rotating in the opposite direction (→ or ←); that is to say, if one tornado is rotating to the right and the other tornado is rotating to the left, these two tornadoes would not integrate; so, the two tornadoes would continue to exist independently.

The sign (±) of the two integrated tornadoes would only indicate in which direction the new tornado is turning.

Just as the positive electronic energy of the 2 positive fermions +1/2 and +1/2 or (↑↑) was integrated spontaneously and the positive electronic matter of the electronic nuclei is formed; in the same way, the electronic energy of the 2 negative fermions -1/2 and -1/2 or (↓↓) is integrated spontaneously and the negative electronic matter of the electrons is formed.

The electronic energy of almatrino 5, i.e. +1/2 (↑) of level 1, is the electronic energy connecting energy level 1 to energy level 2; and so on in a successive manner. In this way, the values of the energy levels went towards a very large value of the electronic energy.

We can write in a more general way as: $+(n_1/2)$, $-(n_1/2)$, $+(n_2/2)$, $-(n_2/2)$, $+(n_3/2)$, $-(n_3/2)$... $\pm(n_n/2)$, and so on.

We cannot write 'n' infinitely, (n_∞) because the formation of the Universe has not ended and cannot end; for, the Universe is expanding into nothingness. So, the expansion of the Universe will not be able to reach an infinite value.

The flow of the electronic energy of the fermions +1/2 (↑) and -/1/2 (↓) generates a magnetic energy. The magnetic energy rotating from left to right (→) is positive; and, the magnetic energy rotating from right to left (←) is negative.

These 2 positive or left-to-right spinning magnetic energies (→→) are integrated and form the magnetic mass of a male being.

The 2 negative or right-to-left magnetic energies (←←) are integrated and form the magnetic mass of a female being.

Because of the movement, the energy flow of the almatrino, being energy, could not remain static. Moreover, the flow of energy is towards nothingness, i.e. there are no forces in nothingness that are able to stop the flow of the energy of the Universe.

For example, when the almatrino was half way through its elliptical orbit, the other elliptical half of the almatrino's path had already increased to an exponential value of the form $y=e^{xt}$. That is, the growth of the Universe is not linear or of the form y=xt; rather, the generation of the energy of the Universe is in an exponential manner. It means that every time the energy of the Universe grows, the Universe expands over an amount of energy that already exists.

At that time, the temperature of the Universe was very high, because the size of the Universe was infinitesimal. So, the almatrino had to rotate around itself faster and faster in order to complete its elliptical orbit. The elliptical orbit of the path was getting bigger and bigger for the almatrino. By moving faster and faster, the almatrino created more energy; the energy increased the temperature and by the expansion, more space was created. But the rotational speed of the almatrino could not be infinite to complete its elliptical path. Therefore, a speed value was reached where the electronic rotational energy was converted into electronic matter.

If this conversion of electronic energy into electronic matter had not happened, the almatrino would still be rotating around itself at an exponentially increasing speed.

The equation that predicts how fast electronic energy is converted into electronic matter is $Ev=m_0C^3$. In this equation, as mentioned, m_0 is the initial amount of electronic matter in the Universe, E is the energy generated, v is the speed of the spin of the almatrino, and C is the constant of proportionality. The proportionality constant C is the one that allows us to introduce the value of equality between the quantities.

However, the conversion of the electronic energy into electronic matter cannot be done from the equation of Albert Einstein and Mileva Marić $E=mC^2$; since, with this energy equation of Albert Einstein and Mileva Marić we do not know if 'm' in the equation is the electronic matter or if 'm' is the magnetic mass. For Albert Einstein the initial electronic matter of the Universe m_0 is imaginary, and Albert Einstein's energy equation did not consider the velocity v of energy.

But, electronic matter is different from magnetic mass. Electronic energy is produced by motion; electronic matter is formed by the integration of electronic energy; whereas, magnetic energy is produced by the flow of electronic energy; and, magnetic mass is formed by the integration of magnetic energy.

The positive electronic matter of the nuclei is spatially integrated with the negative electronic matter of the electrons, and molecules, or the elements of the periodic table, are formed.

In nature, these elements of the periodic table are usually not in a pure state. For example, there is no hydrogen atom H, but a hydrogen molecule H_2, because as a molecule, the two hydrogen atoms dynamically compensate each other's electronic charge.

Thus, through this integration of electronic matter, all the electronic matter in the Universe is formed spatially between the positive nuclei and the negative electrons.

Electronic matter has nothing to do with magnetic mass; but, if we compare the forces of integration, the integrating force of electronic matter is less intense than the integrating force of magnetic mass. For, as mentioned, the electronic matter is spatially integrated between the negative electrons with the positive electronic nuclei. Therefore, the negative electrons can move away from the positive nucleus in the form of electromagnetic radiation. Whereas, the integration of magnetic energy in the form of magnetic mass is not spatial, because the magnetic mass has no nuclei and no electrons.

We call the electronic bosons that integrate electronic matter gluons. While, the bosons that make up the magnetic mass can be called urdires.

The 2 positive magnetic energies, or energies that rotate in the same direction (→→), integrate spontaneously and the positive magnetic mass of a male being is formed.

The two magnetic energies, which rotate from right to left or negative (←←), spontaneously integrate and form the negative magnetic mass of a female being.

Thus in the human race the male and female, or in animals the two genders, the female gender and the male gender, were formed. For example, in other lineages it will be a cock and a hen; a lion and a lioness, a cat and a female cat, a bull and a cow, etc.

These two beings, male and female, can be spatially united by means of the physical bodies. By the spatial union of these two physical bodies, another physical body will be formed; and into this body will be incorporated the spirit formed by the magnetic mass coming from the spiritual world.

The incorporation of the spirit into its new physical body will take place 5 months after gestation. And it will be at 5 months, because the embryo was formed from the physical integration of two haploids. The male haploid is in the testicles of the male being, and the female haploid is in the egg of the female being.

Haploids have no physical memory. Physical memory is weightless magnetic energy; therefore, magnetic memory is embodied in the spirit that is incorporated into a physical body.

The mass and magnetic memory is what gives the style or form of life to a spirit being living in a physical body made up of electronic matter.

The magnetic memory of the spirit sits in physical form in the hippocampus. Therefore, babies have no physical memory; so babies can see and talk to a spirit; because babies can see things that move very fast. Likewise, babies can hear the infra-sonic range. This will happen to the baby, until the baby reaches the age of a 5 year old child. At that age, the formation of the hippocampus will culminate in the child's brain.

According to this sequence (↑↓↑↓↑) we deduce, that the almatrino number 1; that is, the almatrino that began to form the Universe had an electronic energy pointing upwards; or that the magnetic energy of the first almatrino rotated in the right-left direction (←).

Although the spin of the almatrino is relative; since, the direction of the spin is going to depend on which side we are seeing the almatrino spinning. If someone is seeing the almatrino spinning from left to right, the other side will see that the almatrino is spinning from right to left.

In the beginning, the infinitesimal electronic matter of the infinitesimal Universe was integrated with the infinitesimal magnetic mass; and thus, the various infinitesimal physical bodies with the capacity for life were formed.

Some of these infinitesimal bodies live dormant or asleep inside an infinitesimal capsule, just waiting for environmental conditions to awaken from their dormant state and reproduce.

We call these infinitesimal bodies viruses.

These infinitesimal physical bodies awoke in the hostile environment of Earth. It is likely that this began in the desert of Lut, or on a high plateau and from there formed the different kinds of cells that gave rise to living things on Earth. Including vegetation that would form first by the effect of condensation of water vapour, then by the effect of tides that dragged phytoplankton from the sea to the infertile land.

Plants consume the carbon dioxide that animals expel through their noses. Plants discard oxygen as waste. Oxygen is consumed by living things to produce energy in the form of heat. Thus, an ecosystem was formed that gave rise to the relationship and coexistence between the different forms of life on Earth.

In the beginning, the Universe was dark, because the planets and the atmosphere of the planets had not formed to break down electromagnetic radiation into light and heat.

The Universe has grown, because the electronic energy that creates the space of the Universe is of an exponential form. If a human being were to stand at the edge of the Universe to see the rate of growth of the Universe he would not notice it; for, the physical space of the Universe is very large. So, what represents an instant to the Universe will be an eternity to a human observer.

The growth of the Universe is towards nothingness, and it will be chaotic, because in nothingness there is nothing that can stop the growth of the Universe. We owe the idea of chaotic motion to the French scientist Jules Henri Poincaré.

This is part of a constant search to find out how the Universe was formed in a logical way. For example, Georges Lemaître was a priest, but Georges Lemaître did not succeed in demonstrating how the energy that moves the Universe exponentially was formed. Georges Lemaître only developed the Big Bang theory.

However, in order to bring the Universe back to its starting point as Georges Lemaître's Big Bang theory proposes, we would have to decrease the entropy of the Universe. But, it will be impossible to pick up the Universe again; since, in order to pick up the Universe again, we would have to provide it with more energy than the Universe has produced in order to bring the Universe back to its initial point. In other words, the Big Bang theory does not explain where the energy and matter that formed the Universe came from.

Maybe this is an assumption that can be made mathematically, but the possibility of picking up the Universe again does not make sense from a physical point of view. So, we will not be able to contract the Universe to bring it back to its starting point.

At the starting point, we will not be able to have an infinite density of electronic matter; but, using the Big Bang theory, we will also not know where the energy that heated the starting point of the Universe came from.

So, with the analysis we have made, of how all the matter that is forming the Universe and all the matter that will form in the Universe was formed; the magnetic mass of the spirits, and the male and female genders, which in the human race is man and woman, represents the most important scientific fact that has occurred in the entire history of scientific thought in human civilisation.

However, it is easier to understand a philosophy than a scientific fact; but, we cannot senselessly hinder scientific analysis; therefore, we must allow scientific explanation to progress logically; according to the British Francis Bacon and the Italian Galileo Galilei.

So, all human beings are obliged to understand where and how all electronic matter and magnetic energy in the Universe arose and how it will arise, through the energy equation that formed the Universe $Ev=m_0C^3$, in order to make humanity more humane.

Chapter 3

THE EQUATION THAT EXPLAINS HOW THE UNIVERSE WAS FORMED

The expansion process of the Universe is irreversible; therefore, we cannot wait for the Universe to retract by itself to its initial point. It is impossible to go from chaos to order spontaneously, since, in order to reverse disorder to order, energy must be contributed to the system. So, at this point in scientific history, the Big Bang theory of the Belgian reverend, mathematician and astronomer Georges Henry Joseph Édouard Lemaître is finished. The Big Bang theory or the cosmic egg theory does not make logical sense, since it does not explain, for example, where the energy that heated the nascent Universe came from. Although new theories explaining the formation of the Universe have been developed from the Big Bang idea, we will not be able to concentrate all the energy that the Universe has now, and the energy that the Universe will have in infinite time, into one point.

However, again, the explanatory vacuum led many scientists to assume the Big Bang theory, and those derived from it, as the most representative explanation of the formation of the Universe.

With the expansion of the Universe into nothingness, what we have is chaos or disorder; since, what the Universe seeks with the expansion is a state where the energy system that forms the Universe is in thermal equilibrium. That is, so that the Universe can reach a point where it has less energy; for which, it is required to go from order to disorder. But, from this point of view, the point of equilibrium of the Universe is not even beyond the plus infinite; since, the Universe creates by itself the energy that drives it towards nothingness; and in nothingness, there is no energy.

The Universe forms an implosion in an absolute vacuum.

The inner boundary of nothingness is equal to the outer boundary of the Universe. We can define that boundary as a flexible edge; that is, the boundary of nothingness expands as the Universe expands; for, outside the Universe there is nothing. Therefore, the energy flow of the Universe will be eternally towards nothingness, and the entropy or disorder of the Universe will increase continuously towards nothingness.

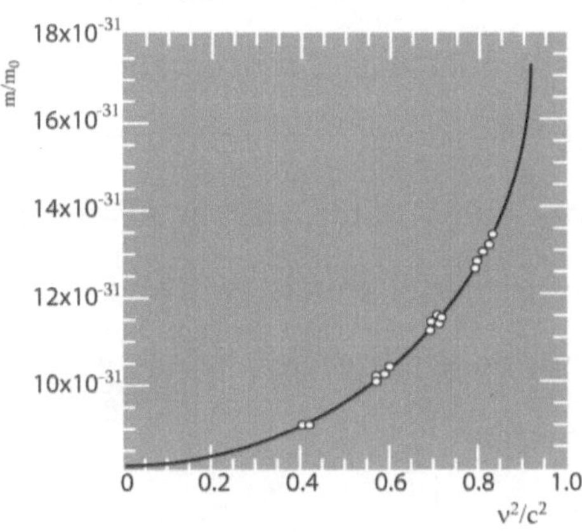

FIGURE 1

BUCHERER AND NEUMANN GRAPH

However, Albert Einstein did not want to see in Bucherer and Newmann's Figure 1, that the speed beyond the value v^2/C^2 goes to an infinite value. When the speed is greater than 1 ($v^2/C^2>1$), Albert Einstein deduced that, if something could move with a speed C greater than light, the mass of this particle would not be real but imaginary. But, in reality it is not like that; since the electronic matter of the system that is forming the Universe in the nothingness is real; because the electronic matter is formed from the electronic energy, and the electronic energy is a real magnitude.

That is to say that, as Ralph Kronig deduced, any real particle, has to rotate on itself with a higher speed than its translation speed, in order to move along its elliptical orbit. Since, the elliptical form is the only way of movement, so that the particle can contain the same amount of energy in its journey through the same energy level. That is, with an elliptical path, the particle does not lose energy; that is, with an elliptical motion, the particle does not decelerate, as it would if the motion of the particle were circular.

The elliptical form for the translation is the spontaneous or natural way of motion. If the translational motion had been circular, the circular paths within a sphere would not have been formed. That is to say, if the motion were circular, all particles would have agglomerated in a single orbit and the space that forms the Universe would not have been created. With a translational motion of energy in a circular orbit, the energy system that forms the Universe would not have expanded; or, the system would not have formed out of nothing. If it had happened in this way, the energy system that forms the Universe would have reached a state of exhaustion; or, the Universe would have the form of an energy belt.

That is, the geometry of the Universe is spherical. If the shape of the Universe had been elliptical, the system would have consumed energy into nothingness, and the Universe would be in equilibrium. That is, the Universe would have reached a state where the entropy value is at a maximum. But, entropy is a spontaneous process that cannot be stopped; therefore, we will not be able to reach a point where entropy has a maximum value. It is impossible to reach a point where entropy has a maximum value; therefore, the Universe will continue to expand progressively towards nothingness.

The energy system of the Universe cannot go backwards, because the Universe would become nothingness at its initial point; but this is impossible, because the Universe cannot go back to its initial point.

The process of the formation of the Universe takes 13.8 billion years, so we will not be able to pick up the Universe again.

We can only pick up the Universe as a theoretical probability by mathematical imagination, but this assumption does not make sense from a physical point of view, because retro traction is not a spontaneous process. It requires more energy to be put into the system than the amount of energy the Universe has produced in order to bring the Universe back to its starting point.

Entropy does not decrease on its own, because that would be like saying that something that was disordered became orderly on its own. There is not enough energy in the Universe to spontaneously bring the Universe back to its starting point.

The space separating the different bodies has to be large; as, for example, the space between the bodies forming the solar system. For, when two bodies rotate in opposite directions (← or →) a repulsive force is formed between two bodies; as in the case of the two tornadoes rotating in opposite directions. These spatial bodies are sequentially ordered by an alternating configuration of the electronic motion, i.e. positive-negative-positive-negative, (↑↓↑↓) ... etc. But, while they are separate, there is an integrating force that holds them together in the positive (↑↑) and negative (↓↓) form.

For example, the Sun rotates from left to right (→), because the flow of the Sun's positive electronic energy is upwards (↑). Mercury rotates from right to left (←), because the flow of Mercury's negative energy is opposite to the flow of the Sun's positive electronic energy (↓). Venus rotates from left to right (→) just like the Sun or in the opposite direction to Mercury; but, the electronic energy of Venus is negative (↓). The Earth rotates from right to left (←); that is, the Earth rotates opposite to Venus; therefore, the positive pole of the Earth is upwards (↑); and the negative pole of the Earth is downwards (↓). Then we can go with the explanation towards the planet Mars.

However, the planets: Mercury, Venus, Earth, Mars, etc., form the negative charge (↓↓↓↓), and between all of them, they compensate the positive charge of the Sun's nucleus (↑); and because it is an electronic nucleus, the Sun is more massive. Thus, the Sun is larger than the sum of all the planets. For example, the negative charge flows from the Earth into the Sun's electronic core.

So, the motion of each of the electronic bodies is opposite to the next; since, the rotational motion of all bodies in space is a transmission of motion. Therefore, the bodies in the space of the Universe are at the same time integrated by electronic forces of opposite rotation at the same energetic level; which prevents the bodies from merging to form a single electronic body. We call this energy of repulsion and attraction between celestial bodies on the same energy level the electronic force of gravity.

Spirits do not contain electronic matter; spirits are only made up of magnetic mass; therefore, the magnetic mass of spirits is not influenced by the electronic force of gravity.

But we have to describe mathematically how an energy system was formed and since it began, it is still growing exponentially immersed in nothingness.

From the energy equation of Albert Einstein and Mileva Marić $E=mC^2$, we can use the complex number derived by Johann Carl Friedrich Gauss, because the equation of Albert Einstein and Mileva Marić shows how energy can be transformed into electronic matter and electronic matter can be converted back into energy.

Electronic matter can be converted back into electronic energy if the electronic matter comes into contact with a black hole to spin at high speed. But, these systems are neither holes nor black holes, they are energetic systems where electronic energy spins with high speed. Black holes are accretion spheres, where electronic energy is converted into electronic matter.

A spirit has no electronic matter; therefore, a spirit is not affected by a black hole; but, a spirit is aware of its existence; so, no spirit would ever think of coming into contact with a black hole.

We have said that there are particles which have neither mass nor electronic charge; but electronic matter, being condensed energy, cannot be imaginary. For example, spirits are real beings; but, spirits do not contain electronic matter. Spirits are made up of magnetic mass that contains no electronic matter, because spirits have no nuclei or electrons.

However, Albert Einstein considered, that the initial matter was not real; rather, the initial matter would be imaginary according to the mathematical deduction of Johann Carl Friedrich Gauss; but, Gauss was referring to a complex value, which is different from the imaginary value; since, the imaginary is not real.

A complex value is one that is made up of several real elements, whereas an imaginary value exists only in the imagination, i.e. the imaginary value has no real physical form. However, the electronic matter that exists in the system that forms the Universe and nothingness must have been formed at some point and at some time. Therefore, it is necessary to look for an equation to show us in a physical way how the initial electronic matter of the energy system that forms the Universe was formed.

We can subtract from the value 1 the value corresponding to v^2/C^2; that is, the value that is within the square root in the energetic equation of Albert Einstein and Mileva Marić, so that the rotation speed has a limit; or so that the electronic matter m0 does not have an imaginary value. The initial electronic matter is determined by the following relation:

$$m=m_0/\sqrt{1-v^2/C^2}$$

From the consideration that v is greater than C, it means that: (v^2/C^2) or that C is less than v; i.e. (v^2/C^2) is greater than 1. Subtracting the value (v^2/C^2) from 1, a negative value will remain within the square root of the form: $-(v^2/C^2)$. We can use the complex number 'i' by Johann Carl Friedrich Gauss to solve for the negative value of the square root.

So that:

$$m=m_0/\sqrt{-v^2/C^2}$$

Substituting the value of this electronic matter 'm' for the value of the initial matter m_0 above in the Einstein equation we have:

$$E=m_0C^3/iv$$

$$iv=m_0C^3/E$$

Where 'i' is the value of the complex number. Let us multiply both sides of the equality by the complex number 'i'; but, bearing in mind that: $i^2=-1$. Therefore:

$$i^2v=i.m_0C^3/E$$

$$-v=i.m_0C^3/E$$

If we raise the moduli of equality to the square; since, the real value of the velocity is increasing, as is the value of the electronic matter and the value of the electronic energy, so these variables have no mathematical sign. We have that:

$$-|v|^2=i^2m_0^2C^6/E^2$$

$$-v^2=-m_0^2C^6/E^2$$

$$v^2=m_0^2C^6/E^2$$

If we extract the value inside the square root, to calculate the velocity v of the first almatrino that became electronic matter m_0, when the Universe had the size of an infinitesimal energetic sphere, we obtain that the speed of the energy inside the system that started to form the Universe from nothing is:

$$v=m_0C^3/E$$

That is, the infinitesimal value of the energy E of nothingness in a fraction of time greater than time zero was equal to:

$$E=m_0C^3/v$$

We say that it was a greater fraction of time after time zero, because the amount of electronic matter 'm' that the Universe had at this moment arose from the initial electronic matter m_0; since, at time zero, the initial matter of the Universe m_0 was zero. The fraction of time measured from the zero point is $5.391x10^{-44}$ seconds, i.e. the minimum Max Planck time.

We can write in an integrated way the equation that formed the Universe as: $\Delta Ev=\Delta mC^3$. $\Delta E=E_f-E_0$, $\Delta m=m_f-m_0$. That is, with this equation that formed the Universe, $Ev=m_0C^3$ we can know, what amount of electronic energy E_f and what amount of electronic matter mf we will have at any moment.

The initial velocity v was really zero, and this velocity v could not go to an infinite value, because the conversion of electronic energy into electronic matter has a limit in the speed of rotation of the electronic energy. That is to say that: $\Delta v=v$; hence, $\Delta Ev=\Delta mC^3$.

The equation can be applied to any energy system. This is the mathematical equation, which explains how the Universe was formed from nothing.

The equation that formed the Universe can be integrated; to know, how much electronic matter the Universe will have at any point in space. Let's say, from point zero to a point beyond infinity.

The initial space of the minimum Universe was smaller than an almatrino; therefore, we have arrived to the instant; that is to say, to the minimum point or the point zero where the Universe began to form. Therefore, all the measurements that we make, we will not be able to do them in a relative way but in an absolute way, starting from the point zero of the Universe; which is in the mere centre of nothingness. And it was from this point zero that the Universe began to form.

So, the equation $Ev=m_0C^3$, is the relation that explains how the Universe started to form from nothing; and how the Universe is still forming; because the process of the formation of the Universe, once started, cannot be stopped.

If the process of the creation of the Universe were to stop at a point, there would be no more motion; therefore, the process of the generation of electronic energy would cease and the Universe would freeze. What maintains the generation of electronic energy in the Universe is the motion of the Universe.

The Universe can best be defined as an accretion sphere.

The value of the electronic energy E, is related to the magnetic energy B, by means of the constant C. That is, E=CB; therefore, we can write for the magnetic energy B:

$$CB=m_0C^3/v$$

The magnetic energy B of the bubble of electronic energy that formed the Universe out of nothing is:

$$B=m_0C^2/v$$

$$Bv=m_0C^2$$

So, considering that at the beginning the velocity v was virtually smaller than zero, we can say that there was nothing there. Therefore, these two equations: $Ev=m_0C^3$ and $Bv=m_0C^2$, allow us to connect in a mathematical way, an almatrino that emerged from nothing with the physical system of the Universe.

Finally we conclude that, the Universe cannot be static; but, the motion of the Universe we do not notice it, because we march in space with zero velocity with respect to the Earth.

If the Universe were static, the Universe would be in thermal equilibrium; and, in this stopped form, in the system that forms absolute nothingness with the Universe, the flow of electronic energy would not exist; that is to say, the electronic energy that forces more movement to take place. And with more movement of electronic energy, more electronic energy will be created; which will have to be converted into electronic matter, so that the system does not heat up infinitely. So, the Universe will always be in motion; and nothing can stop it.

THE AUTHOR'S WORK

Graduated from the School of Chemistry, Faculty of Sciences, Universidad Central de Venezuela, with a degree in Chemical Technology. Postgraduate studies in Food Science and Technology. Special work on the chemistry of natural products and the chemistry of diseases. Chemical process designer. The books listed below are the product of thought; therefore, these books must be subject to revision as we become clearer about how the Universe was formed, so try to read the latest edition of each book. These books are: "The Chemistry of Cancer". "The Chemistry of Diabetes. "The Heart Attack". "Alzheimer's". "The Chemistry of Arthritis". "The Chemistry of Thought". "The Chemistry of the Spirit". "How the Universe was formed". "The Expensalists". "Why You Shouldn't Eat Meat". "The Micro World". "Does God Really Exist?". "Objecting to Albert Einstein's Relativity". "Divining the Future". "The Mistake of the Great Scientists". "Life on the Sun". "The Universe before Zero Time". "The Energy of the Spirit". "The Origin of Cancer". "The World of Cells". "The Chemistry of Disease". "The Particle that Created the Universe". The Chemistry of Cancer, seventh edition. The Chemistry of Diabetes, sixth edition; The Chemistry of Heart Attack, fourth edition; "The Chemistry of Memory"; The Chemistry of Arthritis, third edition. "The Creative Power of the Mind. The

Particle that Formed the Universe, third edition. "The Initial Mass of the Universe". "You Shouldn't Eat Meat". "The Origin of the Body and the Spirit". "Worship the Universe". "Sugar an Enemy in the Kitchen". "Time Travel". The Chemistry of Cancer Edition 8. The Chemistry of Diabetes, Edition 7. The Chemistry of Heart Attack Edition 5. The Memory of the Spirit Edition 1, The Chemistry of Arthritis Edition 5. "The Life of Spirit". "Re-writing Science". "The Beginning of the Universe". "Spiritual Growth". "Coupling of the Spirit with the Body". "The Origin of Life". "Death Does Not Exist". The Chemistry of Cancer, final edition. The Particle that Created the Universe, final edition. "Incorporation of the Spirit into the Physical Body". "I Came from the Sun". "Preventable Diseases". "The Origin of Life on Earth". "The Inaugural Moment of the Universe".

www.ingramcontent.com/pod-product-compliance
Lightning Source LLC
LaVergne TN
LVHW091236150826
845673LV00003B/1157